CAREERS IN GEOLOGY

GEOSCIENCES

GEOLOGY IS THE SCIENTIFIC STUDY OF THE EARTH, its composition, its processes, and the forces that act upon it. It is a broad subject that covers very specific aspects from glaciers and volcanoes, to gem stones and energy resources, to changing land formations and mass extinctions. It includes every area – the earth's core, ocean floor, deep canyons, mountaintops, and even the atmosphere.

Geologists spend most of their time outdoors, often in remote areas. They dig up fossils, take soil samples, create maps, and gather lots of photographic evidence. They study the weather and investigate potential geological activity in order to predict natural disasters and potentially save people from the ravages of tornadoes, earthquakes, tsunamis, or volcanic eruptions.

There are dozens of different jobs that a geologist can hold. Each utilizes the knowledge and skills acquired from the same basic training and education. What any one geologist does depends on the job title or area of specialization. For example, environmental geologists are concerned with the safe use of natural resources. They test soil and water for signs of toxins after accidents, help create plans for cleanup, and make sure areas are safe for residents. Hydrogeologists work primarily with water. They study how water moves, how and where it becomes available to communities, ways to increase water supplies, and how to minimize possible pollution. Petroleum geologists search for sources of oil and gas, and develop methods for safe extraction.

The minimum educational requirement to become a geologist is a bachelor's degree in geology, though many employers prefer a master's degree. In either case, those entering the field can expect to find jobs waiting for them. In fact, industry leaders predict that some areas will experience shortages of trained professionals as the demand for renewable and safe energy, more accurate hazard weather plans, global environmental safety, and answers to the threat of climate change grows in importance.

Some of the many different organizations hiring geologists include:

- Engineering firms

- Environmental research organizations

- Universities

- Federal and military agencies

- Energy consulting agencies International institutes

- Oil and gas companies City planning commissions

Those who work for these different employers do different types of work, from pure historical research to solving industrial problems. Depending on the particular job, you can start at an excellent salary straight out of college, and reach well into the six-figure range within a few years.

Geology is a career for people with all kinds of interests. Do you like playing detective, searching for bits of evidence and putting them together to solve a problem? Geology is perfect for those with an insatiable curiosity and the ability to visualize in 3D. Were you the kid who played with dinosaurs and never got over your fascination for the prehistoric creatures? Geology could be a career dream come true. Are you concerned about the earth's ability to sustain life in light of environmental challenges and global climate change? You could make a difference. Whatever your interests and goals, geology can lead to an exciting and very satisfying career.

WHAT YOU CAN DO NOW

THE HIGH SCHOOL CURRICULUM FOR A FUTURE geologist should be heavy with science classes. Especially important is earth science, biology, chemistry, physics, and computer science. Math is important, including at least two years of algebra and one year each of geometry and trigonometry. Four years of English is also advised. Like all scientists, geologists need to communicate their findings to others regularly. A class in public speaking would also be helpful. Take Advanced Placement classes if available. Check the admission requirements for colleges you are considering to make sure you cover everything.

Start hanging out with people who share your interests. Look for after-school organizations such as science clubs and geology clubs. Find someone who would like to partner up with you on a geology project for the science fair. Geologists often work in remote areas without the benefit of modern conveniences. To get comfortable with roughing it, join an outdoor enthusiasts club and look for opportunities to backpack to places off the beaten track.

Are you wondering what a geologist actually does every day? Ask your school counselor to help arrange some job shadowing. Spending even a single day on the job with a geologist will help you gain a better understanding of what the job is all about and what you need to do to prepare.

College admissions officers and future employers will look favorably on some real world experience in geology. Many colleges and universities offer summer experiences in geology to high school students. These programs range from local field trips to mapping and collecting data from caves and lava fields. Government organizations also offer summer programs and internships. For example, the US Geological Survey (USGS) has high school students interning at sites all over the country – and they even get paid for the work they do.

HISTORY OF THE CAREER

PEOPLE HAVE WONDERED ABOUT THE ORIGIN of the earth for countless centuries. However, the first geological concepts were not developed until the 4th century BC in ancient Greece. It was Aristotle who, observing the composition of land, theorized that physical changes to the earth occurred so slowly that they could not be noticed during a person's lifetime. At about the same time, Roman scientists were noting the difference between rocks and minerals. The Romans became quite skilled at mining specific rocks, such as marble, to use for building the empire.

Greek philosopher and student of Aristotle, Theophrastus, wrote an advanced treatise titled, On Stones. The work described and classified many rocks and gems as well as building materials like limestone and various types of marble. It included instructions on how to assay certain alloys. He also mentioned mining and discussed particular gold, silver, and copper mines that were creating wealth for cities.

On Stones was the go-to reference for identifying minerals until Roman naturalist, Pliny the Elder significantly expanded on the subject in 77 AD. In his encyclopedic Naturalis Historia (Natural History), Pliny described many more minerals and metals, particularly those used for practical purposes.

Like the physical changes to the earth, the science of geology moved slowly for the next 700 years. In the Middle Ages, the most interesting theories were formed in China by naturalist Shen Kuo. The scholar studied soil erosion, silt deposits, sedimentary uplift, and marine fossils deposited a hundred miles away from the ocean. His observations formed the basis of geomorphology. However, his most intriguing theory was of gradual climate change. It arose from the observation of ancient petrified bamboo he found preserved underground in a dry northern climate where bamboo no longer thrived.

Various geological theories abounded throughout the Renaissance – most without any scientific basis or even common sense. Even Leonardo da Vinci (15th century) had something to say on the subject. He debunked the popular story of Noah lifting sea creatures out of the ocean and depositing them on the hillsides. He correctly noted that fossils were embedded in the rocks, not on the surface. He concluded this could not have resulted from a temporary flood, but rather the area must have been an ancient seabed.

Some scientific progress was made during the 17th century, particularly in linking the sedimentary deposits in the oceans to the strata in rock formations. A fierce battle raged between scientists and theologians over the true age of the earth. At that time, there was no way for anyone to know that the earth is 4.5 billion years old. Theologians insisted it was only 6,000, years old while the scientists theorized it was much older. In the midst of the debate, a classically trained scientist, Nicolas Steno, questioned much of the conventional wisdom about the earth's processes. He refuted the notion that fossils grew in the ground as well as most theories about rock formation. His research and conclusions formed the basis of stratigraphy as it is known today, which is why he is considered one of the founders of modern geology.

By the mid-18th century, it became acceptable to study the history of the earth without religious repercussions. This cleared the way for geology to become a distinct field of science. The term "geology" was first used by two Swiss naturalists, but the term was not fully established until it appeared in the French Encyclopedia (Systematic Dictionary of the Sciences, Arts, and Crafts) published beginning in 1751. Once the term was fixed, the value of disseminating knowledge of the field became recognized by educational institutions. The first teaching position specifically for geology was created in the National Museum of Natural History in France.

The drive for economic gain propelled geology to the forefront of popular studies in the 18th century. With many more people studying it in a systematic way, the amount of detailed information about the earth quickly grew. Scholars focused mostly on minerals and mineral ores due to the increasing economic impact of mining on global economies. The more efficiently semiprecious metals and other mineral resources could be identified, the more money could be made by the mining industry.

In the 19th century, the mining industry and Industrial Revolution provided economic incentives for governments to support geological research. The motivation was to acquire practical knowledge of geological data that could be used for profit - but even naturalists like Charles Darwin, interested in pure science, were beneficiaries. Government funding meant better technology and the underwriting to explore distant lands. Several countries funded massive geological surveys that produced geological maps that pinpointed useful minerals throughout vast areas.

Geology took a big leap in the 20th century when the theory called "continental drift" was proposed in 1912. That idea led to other areas of thought, including plate tectonics, sea floor spreading, and paleomagnetism. Although there was not enough hard evidence to support the theory of continental drift until after WWII, it was eventually recognized as the cornerstone of modern geology.

Today, geologists continue the tradition of studying the nature and origin of the earth and the processes that affect its surface features and internal structure. What has changed is the inclusion of the atmosphere, biosphere and hydrosphere. Satellite technology has played a major role in this expansion of geological studies. Wide scope photos taken by satellites in space provide a perspective of the earth that cannot possibly be seen from the surface. Stratigraphic principles have been applied to the distribution of the craters on the moon in a series of satellite missions jointly managed by NASA and the US Geological Survey. Satellite images are also used to map, identify, and correlate rock ty pes throughout vast regions and track the movements of tectonic plates. On a practical level, today's geologists use data collected via satellites to locate new sources of natural energy, monitor the effects of global climate change, and predict possible natural disasters caused by plate shifts.

WHERE YOU WILL WORK

THERE ARE ROUGHLY 40,000 GEOLOGIST at work in the US, making it a relatively small occupation. They are scattered throughout the country, but the bulk of geology jobs are concentrated in California, Colorado, Alaska, and the District of Columbia.

Despite its small size, geology is a surprisingly diverse field. A geologist can find positions with a variety of employers, such as nonprofit organizations, universities, government agencies, private sector businesses, and international organizations. Some geologists choose a particular type of job and employer. Others put their broad knowledge and training to work as a consultant to various organizations.

Historically, the majority of geologists worked for mining companies or oil and gas companies, exploring new resources. While these industries still represent the primary source of geology jobs, the demand for geologists is growing faster in other areas. The most active employers in the public sector include:

Federal agencies, such as NASA, US Forest Service, National Park Service, National Oceanic and Atmospheric Administration (NOAA), Bureau of Land Management, and the US Geological Survey.

US intelligence agencies like the CIA, FBI, and NSA, hire geologists.

State agencies hire geologists to work on highway and transportation planning, water resources and coastal commissions, geological surveys, emergency services, and environmental protection.

Local agencies need geologists to work in public works, water and power departments, and city planning.

In the private sector, most jobs are found in:

- Mining companies Petroleum companies Engineering firms

- Environmental research and consulting firms

- Energy consulting agencies Natural resource companies Publishers

- Legal firms

As the earth's resources are shrinking and climate change becomes a growing threat, various international institutions are turning to geologists for help. The main players include the United Nations, CARE, World Bank, Global Environment Facility, and the Intergovernmental Panel on Climate Change.

Work Environment Geologists are extremely versatile. Depending on their particular jobs, they can be found working anywhere. It could be in a scientific laboratory, a classroom, in deep underground caves, the top of a volcano, the ocean floor, or open ground just about anywhere.

Most geologists work at least part of the time outdoors, but some do more field work than others. Exploration geologists spend the most time outside. Their work takes them to remote areas where they encounter varying terrains and weather changes. Environmental geologists divide their time between doing on-site assessments in the field and preparing reports on a computer in an office. Geologists working in academia spend most of their time indoors, teaching and participating on research teams. They still get outdoors though, to lead field trips, conduct research surveys, and collect samples for research projects.

THE WORK YOU WILL DO

GEOLOGISTS STUDY THE EARTH AND THE PROCESSES that affect it. They are interested in all the natural matter that makes up the earth, such as rocks, minerals, plant fossils, oceans, and the atmosphere, as well as how those elements can be safely and effectively utilized by humans.

A big focus of geology is understanding the changes of the earth over time. Natural processes during the past several million years have shaped the earth and affected the existence of every living thing. The biology of ancestral inhabitants can be understood by studying fossil records. Craters hint at cataclysmic events that led to mass extinction of some species and rock formations explain the movement and growth of different civilizations.

Geology is a very useful science. Geologists can use the physical history of the earth to explain current events and help predict future changes, such as land formation, water levels, and climate change. Their studies of earth processes like earthquakes, landslides, floods, and volcanic eruptions are used to survey land and draw up safe building plans. Geologists explore the materials that make up the earth – metals, minerals, water, natural gas, and oil. In addition to discovering new deposits of useful resources, they devise new ways to extract them cleanly and safely. They also investigate ways to clean up areas that have been negatively impacted and monitor environmental remediation.

The duties of a geologist are a combination of fieldwork, laboratory research, and office work. In the field, they create maps and assess areas of geologic activity. They collect samples of rocks or water and explore below the surface using radar or remote sensing equipment. In the lab, the samples are analyzed with electron microscopes and other technology. Specialized computer programs are used to analyze various data and create models. In the office, details are added to maps and charts to identify formations, patterns, and distribution. Scientific reports are then written to document findings.

The field of geology often crosses over into other disciplines, such as civil engineering, urban planning, climatology, environmental studies, and evolutionary biology. The work itself is not confined to the earth. Planetary geologists study the materials and geological processes of other celestial bodies throughout the universe, such as planets and their moons, asteroids, comets, and meteorites. They may never travel into space, but geologists, not astronauts, will likely discover evidence of other life forms or figure out how the universe was created.

Geologists also delve into the oceans. Some marine geologists conduct undersea land surveys or study the mineral content in search of natural resources. Others go deeper, and study the movement of tectonic plates beneath the sea floor. They use their findings to predict seismic shifts and eruptions of underwater volcanoes that could create deadly tsunamis.

Wide Scope of Geological Work Geology is an incredibly large and varied field. There are three general areas of geology work:

Earth materials, which involves locating, harvesting, and preserving the earth's natural resources.

Earth history, using the geologic record of the earth's evolution to understand how it was formed, how it continues to change, and how those changes will shape the future of the planet and its many species.

Earth processes, especially those that are dangerous to people, such as earthquakes, landslides, floods, and volcanic eruptions.

Within each of these areas, there are many different kinds of geologists with tasks that reflect the goals of their employers and the sector in which they work. Though they work for different places and hold different positions, from environment management to education, they all contribute to the overall science of the earth. Some of the most common specialties are described below.

Environmental Geology

Concerned mostly with soil and water and how these materials are involved with the many different areas of the environment. Geologists test these materials to identify any toxins that need to be cleaned up. They may be called in after a polluting incident or in advance of land development to assess whether an area is safe to build on. The environmental issues they investigate and remediate may be caused by human activity, such as major construction or natural resource extraction. Major geologic events like earthquakes and flooding can also have severe environmental repercussions.

Hydrogeology

Also known as geohydrology, this specialty is concerned specifically with water. Geologists in this field study how water moves from place to place and how communities can increase the supply of water while minimizing pollution. Natural underground water reservoirs are of particular interest. Researchers investigate their capacity to store water, how much is being drawn out for human consumption, and whether the cycles of precipitation are able to keep up with the need for replenishment.

Economic Geology

Previously known as petroleum geology, this is about exploiting the earth's natural resources for economic or industrial purposes. These geologists explore every corner of the earth and seas to find oil, natural gas, coal, and any minerals that can be of value. They analyze geological information to identify possible new sites of deposits, collect rock and sediment samples, and test the samples for the presence of resources. They also estimate the size of new and existing deposits and help develop methods of extraction.

Engineering Geology

Specialists concerned with the planning and construction of buildings and bridges, and placement of roads and landfills. Geological engineers apply their knowledge of roc k strength, stability of slopes, and the mechanics and chemistry of soils to the principles of civil and environmental engineering. Their expertise is generally required for major projects, such as highway planning, environmental cleanups, and the mitigation of natural hazards created by tunnels and dams.

Paleontology

This is the study of ancient life and ecosystems. These geologists analyze layers of rock and soil for clues about prehistoric times. They search for fossils much like archaeologists search for remnants of ancient civilizations. The fossils they find in geological formations provide a record of the evolution of plant and animal life, ancestral climates, and environmental conditions throughout the ages. Known for unearthing dinosaur remains that then go on display in natural history museums, paleontologists can determine why certain species went extinct and how that might affect current species. This is a huge field with many subspecialties, such as oceanic paleontology, paleomagnetism, and paleoclimatology.

Geology careers are often highly specialized, with job titles that relate specifically to the area of expertise. For example, a glacial geologist deals solely with glaciers, or a seismologist is concerned with phenomena related to earthquakes. Some areas of specialization require advanced education. A geophysicist, needs a PhD to research the inner workings of the earth, such as tectonic plate shifts, continental formation, and the synthesis of the physical and fluid properties of materials making up the earth.

Even without advanced training in specialized areas, geologists receive broad training that can be applied in numerous ways outside of fieldwork and lab research. There are educators teaching geology in middle school and high school classes. It takes a PhD to teach at the university level, but only a master's or bachelor's degree in some cases to join a pre-college faculty. Geologists with a knack for English or journalism can be science writers or editors of scientific articles, manuals, and other publications. Those with a passion for the environment can use their knowledge of geology to affect public policy or get involved in environmental law.

GEOLOGISTS TELL THEIR OWN STORIES

I Am a Geology Professor

"After 20 years in the field, I came to realize that so much of what I came across while doing my job excited me, yet I had no one to share that excitement with. At least no one who would appreciate my level of enthusiasm. My solution was to go into teaching. Now I have a captive audience with whom I can share my experiences and my love of geology. So far, I've never had a student drop my class so I must be doing something right.

I think my students are special. Majoring in geology is hard. At the beginning of every semester, I let my students know what to expect – nonstop studying and long hard hours in the field. Their typical reaction is, 'bring it on.' I give them advice on what classes are most important, which is often different from what they've been told by admissions officers and counselors. For example, they're often told they need advanced math skills. Truth is, if you're not already good at math there are plenty of areas of geology that do not involve much math. Plus, computer software takes care of most of the equations anyway. So instead of sweating over another perplexing math class, I suggest taking some art classes. It'll make mapping a little easier and help you to develop the ability to visualize in three dimensions. I also advise taking extra communications classes to learn how to write clearly and concisely. It will come in handy when writing abstracts and a senior thesis.

Geology students tend to be a close-knit group, doing a lot of things together both inside and outside of class. Mine is like that. We are all into hiking and camping, which makes field labs a lot of fun. I lead a structural mineralogy lab where we learn things by playing in the mud. I try to nurture their sense of wonder and innate curiosity. I show them that a lot of the work is like being a detective searching for bits of evidence and piecing them together to create the big picture. After a few of my field labs, they never look at a landscape in the same way again. They will forever notice the rocks first and everything else second."

I Am an Economic Geologist

"As an undergraduate geology major, I experienced three different kinds of geology work through internships. First, in the petroleum industry was a 'mud logger' job where I took rock samples from an oil well site and kept a log of what the drills were going through. Next, was hydrogeology. I assisted a team that cored and tested water levels and quality 3,000 feet down. It was fun seeing all the fossils on the cores. The last internship had me exploring for gold and silver on the West Coast. It was like a modern day treasure hunt with cool tools and lots of hiking and camping. After the first 'find' I was hooked. I continued on to complete my master's degree and went to work in mineral exploration.

Economic geology simply means looking for anything of value under the earth's surface. That can cover a lot of materials, including oil and gas, but I specialize in mineral deposits. My work is extremely diversified, so I'm never bored. It takes me to all kinds of amazing and beautiful places. I've worked in many states and countries like Australia, South Africa, and Bolivia.

I spend about half of my time in the field, mostly in the mountains. I map the geology and collect samples that provide clues to the location of mineral deposits. When a potential site is discovered, a drill rig is built to bore holes and gather more data. I use the data to create a three-dimensional model of the deposits.

I love being in the field, but it can be uncomfortable and exhausting at times. I enjoy the challenge, but I'm always careful to avoid dehydration, hypothermia, and injuries from falling. Many places are far from a town where help can be found. When I'm not in the field, I'm in a lab analyzing data.

That can be tedious, but that's about the worst I can say.

Overall, this is a fascinating career that pays me well to find earth's hidden treasures."

PERSONAL QUALIFICATIONS

GEOLOGISTS LOVE ROCKS. THEY ARE INTRIGUED by the mysteries of volcanoes and glaciers. They want to know where precious gems and useful metals are hiding. They study the history behind rock formations above and below ground with wonder. Geologists are naturally drawn to the science that deals with our planet's physical structure and the processes that act on it. An interest in geology, however strong, is not enough to build a career. There are several other personality traits and skills that successful geologists share.

Outdoor Skills

Geology is not one of those stuffy sciences being pursued in sterile laboratories. In fact, geology is dirty work that is most often conducted outdoors. If you have enjoyed camping with your family, friends, or scout troop, you might be ready for the rigors of the job. Geologists often travel to remote areas where camping skills are vital to survival. Do you know how to pitch a tent in the dark, purify water to drink, and build a fire on a windy night? You should also be able to read a topographical map, know basic first aid and CPR, and generally feel comfortable being outside for extended periods of time.

Physical Fitness

Geology involves extensive fieldwork. That does not necessarily mean traveling to faraway places, but it always means physical fitness is required. Some kinds of geology jobs are more physically demanding than others. A volcanologist or seismologist, for example, might need to climb mountains or hike for miles while carrying supplies and tools. Environmental geologists often work closer to home, but they still lug testing and sampling equipment to work sites. In addition to being in good shape, you need to be ready for nasty weather and other uncomfortable outdoor conditions.

Communications Skills

Geologists do more than dig in the dirt. They write everything from abstracts, to field survey reports, to full-length research papers. Sometimes geologists are communicating with other scientists or writing a paper for publication in a professional journal. In that case, scientific writing is the norm. Other times, they are seeking funding from private companies or government organizations. Writing grant applications or proposals is a special skill that requires language that financiers can use to make decisions. There are many reports geologists write to present findings to clients or government officials who do not have a scientific background. For them, clear and concise English is key. Public speaking is also a part of the job for some geologists. Being able to effectively present information to large association, corporate, or government agency meetings can provide a big boost to a geologist's career.

Observation

Geologists are particularly observant people. They are able to notice small key features while capturing the big picture. They are able to visualize in 3D. A keen sense of observation is essential while collecting samples in the field, mapping areas, or charting the composition of certain sites. Good geologists are able to take in information quickly and make accurate judgments. Initial observations are important to determine if and how further investigations should be conducted.

Critical Thinking Skills

Geologists are like detectives. They conduct investigations to solve problems. They gather clues, collect evidence, make observations, weigh the significance of data, and use strong problem-solving skills to arrive at the most logical conclusions. Strong problem solving skills are fundamental to any kind of scientific research, but geology is unusually challenging. Hard facts are not as common as theories, and investigations of ten turn up more questions than answers. Critical thinking skills can help lead to meaningful analyses. Lateral thinking skills can add a creative twist to findings.

ATTRACTIVE FEATURES

GEOLOGISTS REPORT THEY LOVE THEIR WORK because it allows them to follow their passion. It also satisfies their taste for adventure as they explore the mysteries of the earth we all live on. There are advantages and disadvantages to any career, but in the case of geology, the upside seems to far outweigh the downside. Here are the top reasons why:

Geologists earn good salaries that are well above the national average. Starting salaries for beginners can be twice that of graduates with other majors. With a little experience, it is reasonable to expect a six-figure salary in most sectors. Plus, employers often sweeten the pot with a combination of cash bonuses, comprehensive benefits, and other perks.

The job outlook is also excellent. Industry experts predict that in the near future, there will be more jobs available than geology graduates to fill them. This is great news for those seeking entry-level jobs, advancement opportunities, or a stable career in general.

You get to be outdoors. Geology is not like a typical science that is restricted to theoretical bookwork, and you do not have to be stuck in a lab. If you like to hike and camp, look into becoming a field or exploration geologist. Even geologists working as teachers and research scientists, spend half their time out in the field.

Do you like to travel? Geologists tend to travel quite a bit. They may collect samples in less exotic places like Texas or North Dakota, or they might find themselves on an offshore rig or viewing remote fields in the Middle East. Attending conferences is also part of the job. Geologists get to work in a diverse, international environment where they meet and exchange ideas with experts from around the world.

A bachelor's degree can get you in the door. If you are anxious to get started on your new career, there are entry-level jobs that will allow you to do just that. You will want to go on to earn your master's or doctorate at some point though, because an advanced degree will open up many more doors. Plus, as a graduate student you can choose a specialized area of study such as mineralogy, volcanology, hydrology, or even paleontology.

You can be a hero. Many geologists study areas that are prone to flooding, mudslides, and sinkholes, or look for ways to better predict volcanic eruptions and tsunamis. Geologists have saved many lives by helping people prepare for these kinds of natural disasters.

UNATTRACTIVE ASPECTS

FUTURE GEOLOGISTS SHOULD BE PREPARED for hard work, long hours, and employers wanting to see more experience than you have on your résumé. If you are opposed to heavy lifting, blisters on your feet, and at least one sore muscle or bruise on your body on any given day, this may not be the career for you.

Most geologists really like fieldwork. Hiking and camping can be fun, but are you ready to live in harsh conditions far from civilization? Fieldwork also means you will be away from home for extended periods of time. That may not be a problem for new graduates who are young and single. However, it is not conducive to maintaining social relationships or starting a family.

The work can be dangerous. You may be working near deep holes in the ground that penetrate pressurized flammable gases and liquids. As intriguing as volcanoes are, the associated risks should not need explanation.

Some of the most interesting positions require a master's or PhD to compete. In some states, there may be additional requirements such as a state license.

There are disadvantages specific to the type of work. Some of the highest paying jobs for newcomers are in the oil and gas industry. The money is certainly attractive, but job stability

is not part of the deal. Geology jobs are directly affected by international oil prices that can lead to layoffs and hiring freezes.

Working for regulatory agencies often means good pay, even better benefits, and the opportunity for upward mobility. The higher you go, the farther you get from the fun stuff. You will get a big paycheck for spending your days in meetings and at your desk, dealing with non-technical office work.

Consulting firms are taking over many government contracts these days. That means more job openings, but many will take you away from the work you love. You may be assigned project management, proposal writing, and marketing tasks. It is also not a place for introverts since you will be interacting with people more than rocks.

Researchers (and teachers who are required to conduct some research) often get to work on some very interesting projects. However, those projects have to get funded somehow. Writing grant proposals is not nearly as exciting as conducting the projects.

EDUCATION AND TRAINING

A BACHELOR'S DEGREE IN GEOLOGY can lead directly to a career in geology, but it should be considered the minimum requirement for entry-level geology jobs. In many cases, employers will expect you to have a master's degree in geology. For advanced research and jobs in academia, a PhD is needed.

In general, undergraduate geology programs offer excellent preparation for a wide range of career options. Students can choose from several majors:

- General geology Engineering geology Environmental geology Geophysics

- Earth and space science

Each of these provides a deep foundation in geology with courses in mineralogy, petrology, and structural geology. These subjects are important for all geologists. Foundation courses include physical sciences, stratigraphy, geodynamics, and mathematics (calculus). Computer knowledge is particularly important for geologists. Geology majors receive training with the various computer programs specific to geology work. By the time they graduate, they are comfortable working through complex numerical calculations and sets of objective data to reach the most logical solutions. All geology students take classes designed to develop writing and oral presentation skills.

Majors other than general geology focus more on subjects specific to that career track. For example, engineering geology classes would include topics like fluid mechanics, soil mechanics, and strength of materials, hydrology, and surface processes. Environmental geology would involve a wider variety of sciences and include specific courses such as environmental chemistry, microbiology, geographical information systems, and geomorphology. Companies in the oil and gas industry prefer engineering geologists.

The study of geology is unlike any other science major. It is less structured, and students are often confronted with unclear answers and several possible solutions. That is because the work itself deals with messy situations where data collection typically includes a mixture of unclear descriptions and approximated numerical data that leads to open-ended problems. The goal of any good geology professor is to instill students with the ability to use both critical and lateral thinking skills to arrive at the best explanation that decision-makers can utilize in business, government, and public policy.

Gaining Experience

In addition to classroom instruction, geology training takes place in the laboratory and in the field. However, the fieldwork done as part of the regular curriculum should be considered a starting point. Job candidates with the most experience working in the field and in laboratories have the advantage. Most schools offer summer field camp programs where undergraduates can practice putting their classroom knowledge to work. The programs are often led by professors, but professional geologists are sometimes part of the team. Students have an opportunity to perfect their skills in geologic mapping while gaining valuable experience in data collection, two core fundamentals of geology work.

Graduate students in geology have many more opportunities to gain experience. The three most common options are research assistantships (RA), teaching assistantships (TA), and fellowships. Candidates for any of these compete for acceptance. Depending on the type of award, funding may support only educational expenses, although some awards support both educational and living expenses.

Research Assistantships

Professors apply for grants to pay for their research assistants and therefore, they get a lot of say in who that assistant will be. The best strategy for a grad student wanting this kind of assignment is to choose a particular type of research, then contact professors involved in that kind of research at the various universities around the country. Once you have identified opportunities, it is important that you contact each professor directly and make a case for yourself. You can assume that all applicants are going to have knowledge and skills equal to yours. Your advantage will be enthusiasm for the research subject and the opportunity to work closely with the project director (the professor).

Teaching Assistantships

Universities routinely hire graduate teaching assistants to support their faculty. Most teaching assistants act as instructors in introductory laboratories or help with grading. Those with strong skills in particular areas may be asked to assist in one of the upper-level courses or to take over the classroom instruction of a foundation class. To become a TA, applicants must have a broad background in the geological sciences and a solid GPA. In addition, they must demonstrate an aptitude for teaching and have excellent English speaking skills.

Graduate Fellowships

Fellowships are different from assistantships in that they are not connected to a specific work assignment. Their purpose is to encourage promising students to pursue graduate studies in geology. They are much like undergraduate scholarships, covering tuition, fees, and living expenses. Fellowships are awarded to the best of the best students through a highly competitive process. Candidates cannot apply for fellowships. They must be nominated. Decisions are generally based on outstanding academic achievement with strong scores in the Graduate Record Exam (GRE).

Licenses

Some states require geologists to obtain a license to practice. This usually applies to those who offer their services to the public. Requirements vary by state but typically include minimum education and experience requirements, and demonstrating necessary knowledge and skills by passing one or more exams.

EARNINGS

IT IS DIFFICULT TO ESTABLISH AN AVERAGE SALARY for all geologists. Government statistics do not track geologist earnings specifically and in this diverse field, each situation is different. However, the overall average for all geologists is estimated at about $90,000 per year with starting salaries for entry-level geologists ranging from $45,000 to $65,000. Depending on the particular job, a new graduate might start out with a modest salary, but there is plenty of opportunity to advance into higher-paying positions. Top geologists in the right sector in the right location with a graduate degree can earn as much as $250,000 a year.

A geologist's salary is typically negotiated during the hiring process, and varies depending on a combination of factors: experience, industry sector, level of expertise, duties, location, competition, and economic climate.

Experience

The typical salary tends to rise at an accelerated rate throughout the first 20 years. It really takes off during the first five years and continues to increase significantly for another five. The rate of increase starts to slow down at year 10, but does not flatten out until you get near the 20 year mark. That does not mean, however, that senior geologists do not experience additional increases in compensation. On the contrary, the pay structure typically changes to a set base salary, but with cash earnings added on top. Annual bonuses may range from $20,000 to $40,000 and profit sharing, depending on individual performance, can generally mean $25,000 to $50,000 additional income. Other non-cash perks may be part of the overall compensation package as well.

Industry Sector

Geology salaries vary greatly by the type of employer. The highest wages are paid to those in the petroleum industry, where new geologists earn an average of about $80,000 annually. Compare that to the pay in the environmental and government sectors. These employers tend to pay 10 to 40 percent less because they are not in an economic-driven market. There is a trade-off to c on sider though. The petroleum industry is subject to economic fluctuations. Falling prices of oil and gas can result in layoffs and a tighter job market. On the other hand, employment in the environmental and government sectors is often more stable, plus benefits for government employees are exceptional.

Expertise

In this field, expertise is defined as a combination of academic credentials and field experience. Academic credentials are easy to measure – bachelor's, master's, and doctorate degree. The salary structure in geology simply correlates to the level of education completed. Unlike some highly competitive careers, there is no deference given to prominent schools. A geologist with a master's degree from a state university will be offered the same salary as someone with a master's degree from Yale. The only factor that will make a difference is exceptional field experience.

Duties

Job duties in geology can be sorted into three general areas: fieldwork, research, and teaching. Fieldwork is the highest paying job for a geologist, both initially and over the long term. The downside is the long hours and extensive travel, for which there is no additional compensation. The highest paying jobs often come with a degree of risk. Academic jobs often pay more initially, which is why there is so much competition for teaching positions. Unlike most other geology jobs however, they experience a slower rate of increase over time. Salaries for research positions vary widely by type of employer – government backed or independent institute. Government research positions tend to be more stable whereas independent research projects rely on grants that usually come with time limits. Grants can also dry up in times of economic downturns.

Location

Geologists can earn more working in some locations than in others. Top paying jobs in the petroleum industry are naturally found in the parts of the country that are the most dense with oil reserves. Geologists in other sectors will find the highest paying jobs in big cities, particularly those in California, such as San Jose, Sacramento, San Francisco, and Los Angeles. Denver and the District of Columbia are also good choices. If you do not want to live in a city there are also lucrative positions to be found in rural Nevada, southwest Mississippi, the north coast of California, southwestern Alaska, and northwestern Texas.

OPPORTUNITIES

EMPLOYMENT OPPORTUNITIES FOR GEOLOGISTS are very good and there will certainly be no shortage of jobs for new geology graduates in the coming years. In fact, the number of geology job openings is expected to exceed the number of students graduating from university geology programs over the next several years.

There are good reasons the employment rate for geologists is growing faster than other occupations. The field has been steadily gaining more recognition as a tool of the economy rather than for academic research. The demand for new, safe, and renewable energy sources is skyrocketing. Environmental safety and protection issues have risen to the forefront both here in the US and around the world. Growing populations and shrinking resources require more responsible and efficient land and resource management. Plus, the public is clamoring for natural disaster mitigation, particularly in light of weather hazards resulting from climate change.

Government projections indicate that most geology graduates will acquire their first job in industry. Those with a master's degree who have gained some experience working in the field and in laboratories during their schooling will be first in line for the positions of their choice. Employers will look favorably on those with proven skills in computer modeling, data analysis, and digital mapping because they are the best prepared to start work immediately. In addition to a strong academic background, good opportunities await those who are willing to move to cities where major employers are located. The best opportunities are in Denver, Anchorage, and California, especially the southern California cities. Washington DC is also home to many geologists, mostly due to the large number of government agencies that invest in geological studies for the public good.

A graduate degree is certainly more valuable, particularly in terms of earnings, but there is a downside in certain sectors. PhD holders still enjoy easy access to top jobs in the petroleum industry, but may find difficulty landing advanced jobs in research and academia. Those jobs are shrinking in numbers, creating more competition.

The demand for qualified new geologists is stronger in some sectors. The greatest need currently is to increase natural energy output. The number of jobs is outpacing the number of qualified geologists who can help find and harness geothermal energy. Engineering geologists, in particular, are sought after to plan the construction of geothermal power plants. Even though this is a natural and sustainable energy source, it cannot be tapped randomly. A primary task for geologists is to study potential areas to determine how harvesting geothermal energy would affect wildlife, the environment, and nearby populations.

Hydrogeology is another fast-growing sector. There are more jobs than qualified candidates in weather and climate research, storm surge control, and the design of environmental remediation systems. These geologists can also potentially save lives by predicting various natural disasters such as tornadoes, earthquakes, tsunamis, and volcanic activity.

Historically, jobs in oil and gas exploration have dominated the industry. However, employment has always been cyclical and dependent on commodity pricing and government policies. Geologists have played a leading role in the development of technologies that more efficiently access natural gas resources. As more natural gas is produced, prices drop for oil, and the demand for geologists involved in petroleum drops. The future is favoring geologists who can introduce alternative ways to provide energy for the expanding population.

Fewer opportunities are expected in state and federal governments than in the past. Federal agencies like the US Geological Survey, Bureau of Land Management, National Park Service, Environmental Protection Agency, and the Department of Energy are still the biggest employers of geologists. But the overall trend among state and federal agencies is to outsource work to consulting firms. That does not mean that the overall number of jobs is decreasing. The jobs are simply shifting to another type of employer that candidates will need to identify. Fortunately, these employers routinely send recruiters to university campuses to identify the best and brightest candidates. To make sure they can fulfill their government contracts, consulting firms often make offers to students as much as a year in advance of graduation.

GETTING STARTED

THE GEOLOGY COMMUNITY IS SMALL, which presents both challenges and opportunities for the first-time job seeker. The challenge is a limited number of entry-level jobs creating competition for the best positions. The best way to meet this challenge is to get a good education and some hands-on experience. The opportunity is easier to make good contacts in a small field. Aside from education and experience, contacts are your most valuable assets.

Networking is the best way to find geology jobs. Contacts of all kinds can provide job leads. Start building your list of contacts early and do not leave anyone out. Begin with professors and add supervisors and other people you work with in volunteer jobs, internships, and work-study programs. Make more connections by attending geology trade shows, conferences, and seminars. Professional conferences offer excellent networking opportunities for the undergrad and the grad student. It is not unusual to receive tentative offers at these gatherings.

Get your foot in the door through internships. These programs have two purposes. One is to provide real world experience. The other is the possibility of being offered a full-time job when it is over. Expect some competition getting into the most popular internship programs. You can increase your chances by making sure you meet application deadlines and have multiple letters of recommendation from your professors and supervisors of volunteer programs. Do not limit yourself to just one internship. Employers like to see several years of hands-on experience and the odds of a program leading to a permanent job go up every time you participate.

Volunteer positions can also provide good experience for your résumé. It is less likely that you will be offered permanent, paid employment as a result, but you can meet influential contacts that can help you reach career goals. Many government agencies and nonprofit organizations rely heavily on volunteers. Look for volunteer positions at science summer camps and science museums.

Beyond networking, there are numerous ways to find job opportunities. Start with regular visits to your school's career center. This is a great resource for summer jobs, cooperative work programs, job openings, and notices of recruiters coming to campus. Ask to be included on career network emails that the school sends out. After graduation, ask to be added to the alumni network. Most schools continue to send job postings to graduates.

Go where the professionals are. Join professional organizations. Your membership will give you access to job openings and invitations to local chapter meetings. In some of the major industries, such as petroleum, these meetings include dinner and a formal lecture. It is expected that attendees will take advantage of the opportunities to make the acquaintance of other members. It is not considered rude to ask if they know of any job openings (or internships), but do not use this as an opener. It takes a little practice to smoothly work it into the conversation at the appropriate time.

Make a habit of reading updates from professional organizations and science journals regularly. General geology associations are fine, but if you have a specialty, focus on that.

For example, the American Institute of Professional Geologists (AIPG) website has a good job board with positions for all kinds of geologists. If your passion is protecting the environment, you might find your dream job in the jobs section of the International Association for Environmental Hydrology (IAEH) website.

There are also websites not connected to any organization that offer employment listings, such as Geology.com. This site's job board allows you to browse all job offerings or search by industry, type of employer, or specialization. You can also use online job board sites and offline employment agencies that specialize in science careers.

ASSOCIATIONS

- American Geosciences Institute
  www.americangeosciences.org

- Geological Society of America
  www.geosociety.org

- American Institute of Professional Geologists (AIPG)
  www.aipg.org

- American Association of Petroleum Geologists
  www.aapg.org

- International Association for Environmental Hydrology
  hydroweb.com

PERIODICALS

- Geological Magazine
  http://geolmag.geoscienceworld.org

WEBSITES

- US Geological Survey
  www2.usgs.gov/ohr

- Paleontological Research Institute
  www.priweb.org

CAREERS INTERNET DATABASE

www.careers-internet.org